Penguin, Prairie Dog, and Other Colonies

Kylie Burns

A Crabtree Crown Book

Crabtree Publishing
crabtreebooks.com

Author: Kylie Burns

Series research and development:
Reagan Miller, Janine Deschenes

Editor: Janine Deschenes

Proofreader: Kathy Middleton

Design and photo research:
Katherine Berti

Print and production coordinator:
Katherine Berti

Image credits:
Wikimedia Commons
 Pavel Kirillov: p. 27
All other images from Shutterstock

Crabtree Publishing

crabtreebooks.com   800-387-7650

Copyright © 2023 Crabtree Publishing

All rights reserved. No part of this publication may be reproduced, stored in a retrieval system or be transmitted in any form or by any means, electronic, mechanical, photocopying, recording, or otherwise, without the prior written permission of Crabtree Publishing. In Canada: We acknowledge the financial support of the Government of Canada through the Canada Book Fund for our publishing activities.

Hardcover	978-1-0398-0654-2
Paperback	978-1-0398-0680-1
Ebook (pdf)	978-1-0398-0705-1
Epub	978-1-0398-0732-7

Published in Canada
Crabtree Publishing
616 Welland Avenue
St. Catharines, Ontario
L2M 5V6

Published in the United States
Crabtree Publishing
347 Fifth Avenue
Suite 1402-145
New York, New York, 10016

Library and Archives Canada
Cataloguing in Publication
Available at Library and Archives Canada

Library of Congress
Cataloging-in-Publication Data
Available at the Library of Congress

Printed in the U.S.A./012023/CG20220815

Contents

4	What Is a Colony?
6	Why Do Animals Live in Colonies?
8	Penguins
10	Staying Safe
12	Great Blue Herons
14	Flamingos
15	Puffins
16	Prairie Dogs
18	Town Criers
20	California Sea Lions
22	Bats
23	Naked Mole Rats
24	Honeybees
26	Ants
28	Green Sea Turtles
29	Portuguese Man O'Wars
30	Glossary
31	Learning More
32	Index
32	About the Author

What Is a Colony?

Have you ever noticed huge groups of animals gathered in one place? Many animals live together in different kinds of groups. A colony is a large group of animals of the same **species**. They work together to find food, raise young, and protect themselves from **predators**. Living in a colony is like being part of a big family.

*Some animals live in colonies all of the time. Others form **temporary** colonies. For example, great white egrets form colonies only when it is time to **breed**.*

Colonies of emperor penguins huddle together for protection against the bitter cold in Antarctica.

There are many kinds of animal colonies. In insect colonies, each member has a job to do. Many types of birds **nest** together in breeding colonies. Certain **mammals** also live in colonies to help each other survive.

Some colonies have hundreds, thousands, or even millions of members!

Why Do Animals Live in Colonies?

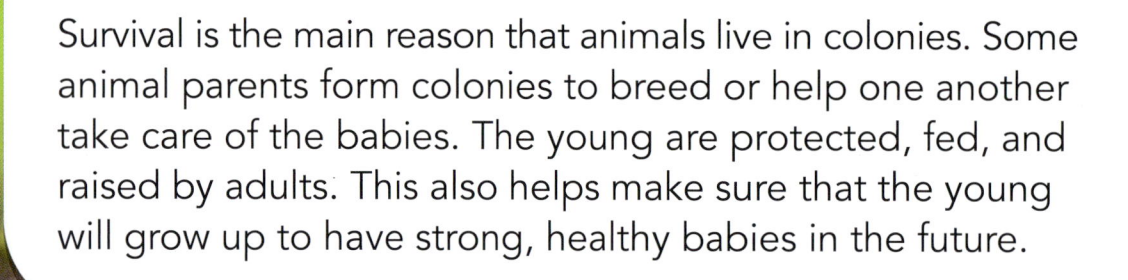

Survival is the main reason that animals live in colonies. Some animal parents form colonies to breed or help one another take care of the babies. The young are protected, fed, and raised by adults. This also helps make sure that the young will grow up to have strong, healthy babies in the future.

Living in a colony helps this Arctic tern baby survive.

Did You Know?

Ants leave a scent trail for other ants to find food. When one ant finds a food source, it marks a path with a special scent. The other members of the colony follow the scent to find the food.

Ants carry food back to the nest for the rest of the colony to eat.

Colonies help animals survive in other ways, too. Monarch butterflies **migrate** south for the winter. At night, they gather in temporary colonies high in the trees so they can rest and keep each other warm. Great blue heron colonies build nests in trees near water. This helps them find food and defend against predators.

Monarch butterfly colonies are so large they sometimes bend or break the branches of trees!

Penguins

At a Glance

Number of species: 18, including emperor, African, and macaroni
Colony size: Dozens to more than one million
Location: Antarctica, Africa, Australia, New Zealand, and South America

Penguins spend most of their lives in cold water. They form breeding colonies on land to lay their eggs and raise their young. These colonies are also called rookeries. Penguins in a rookery work together to stay warm and protect eggs. They also hunt for food together.

Being part of a penguin colony provides protection from danger. Penguins in a colony also have an easier time finding a **mate**. It is easier to care for chicks with the help of others in the colony.

In a colony, adult penguins take turns feeding and protecting the chicks.

Sticking Together

Many penguin parents work as a team to protect their eggs. One parent stays with the egg while the other parent goes to the sea to find food. They continue taking turns to care for the chick after it is born.

Some penguin species carry their eggs in special **brood pouches** above their feet. Others make nests on the ground or build **burrows**.

Penguin eggs have thick shells to protect them from breaking if they accidentally fall onto the ice or rocky ground.

Staying Safe

A lone penguin is an easy target for a hungry predator, such as a leopard seal. Feeding with a group is safer. The penguins stay together to fish. They gather on floating ice when they need to rest. If a predator approaches, penguins flap their flippers and squawk loudly to warn the colony.

These gentoo penguins rest together on floating ice.

A baby penguin's fluffy feathers aren't waterproof, so they can't go with their parents to the ocean to find food. Instead, the chicks huddle together in huge nursery groups called crèches. One adult penguin watches over the crèche until the parents return.

Did You Know?

A colony on the South Sandwich Islands in Antarctica holds the record for the largest penguin colony on Earth. It has around two million chinstrap penguins!

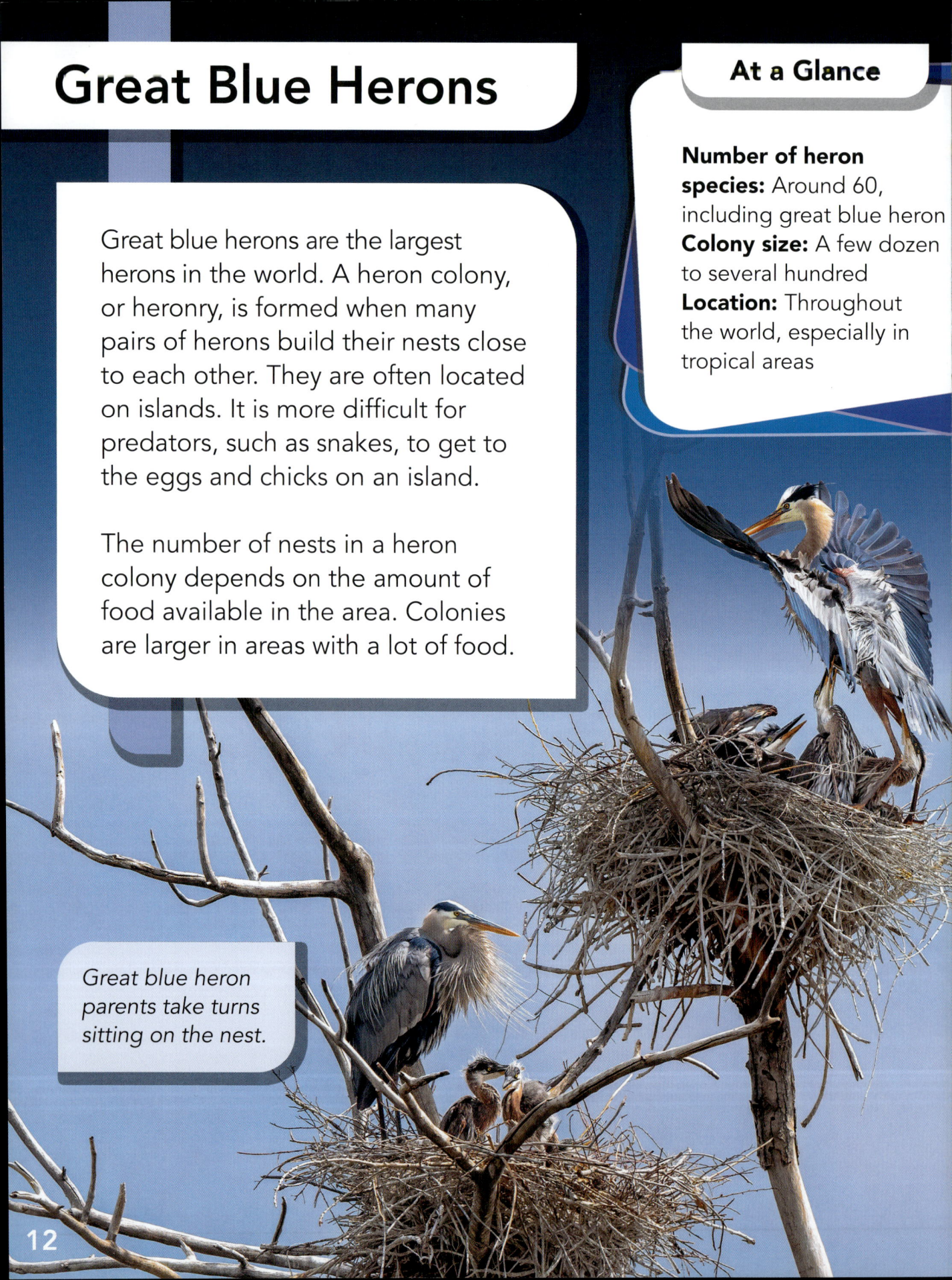

Great Blue Herons

At a Glance

Number of heron species: Around 60, including great blue heron
Colony size: A few dozen to several hundred
Location: Throughout the world, especially in tropical areas

Great blue herons are the largest herons in the world. A heron colony, or heronry, is formed when many pairs of herons build their nests close to each other. They are often located on islands. It is more difficult for predators, such as snakes, to get to the eggs and chicks on an island.

The number of nests in a heron colony depends on the amount of food available in the area. Colonies are larger in areas with a lot of food.

Great blue heron parents take turns sitting on the nest.

Sticking Together

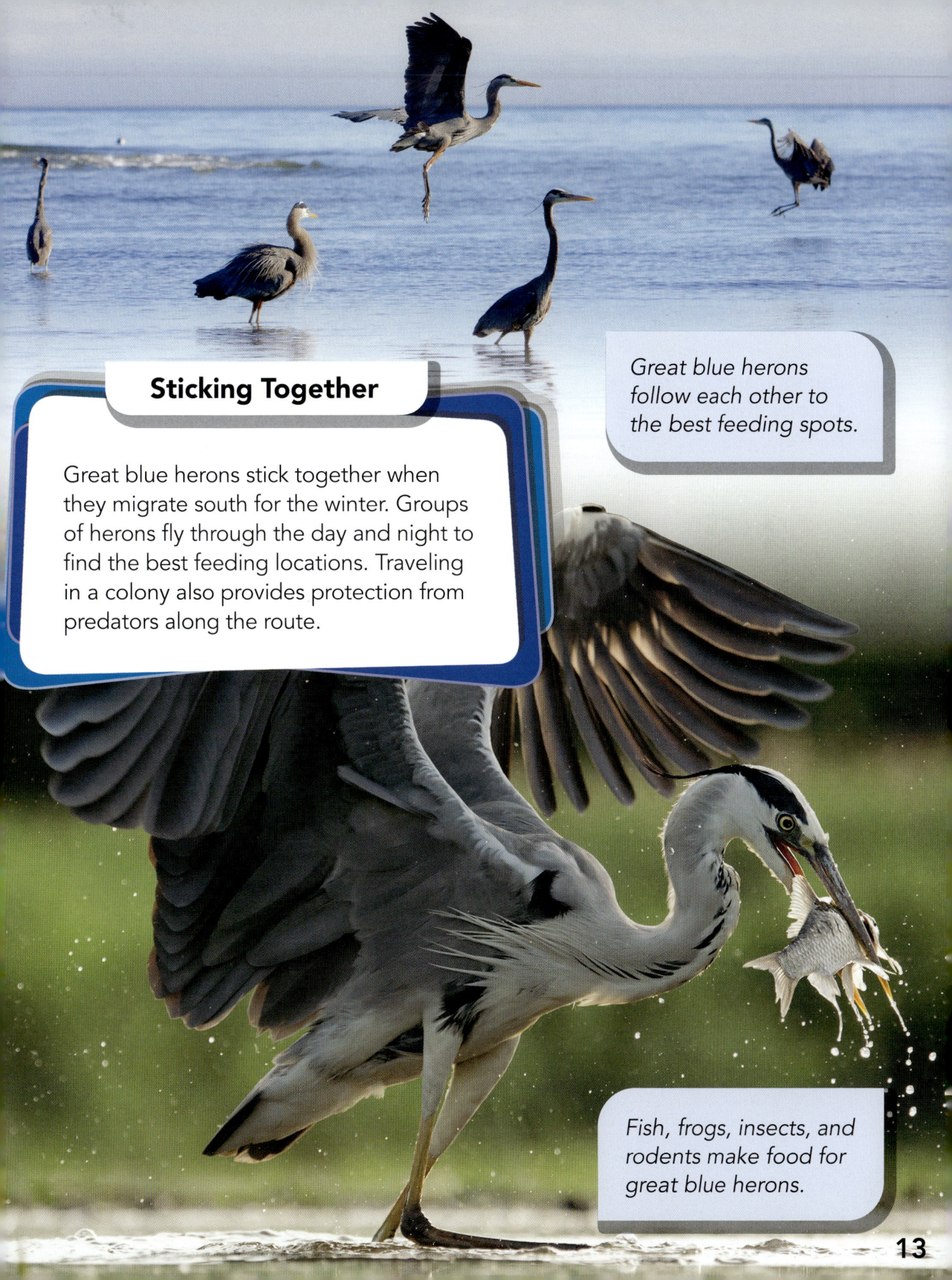

Great blue herons stick together when they migrate south for the winter. Groups of herons fly through the day and night to find the best feeding locations. Traveling in a colony also provides protection from predators along the route.

Great blue herons follow each other to the best feeding spots.

Fish, frogs, insects, and rodents make food for great blue herons.

Flamingos

At a Glance

Number of species: Six—greater, lesser, American, Chilean, James's, and Andean
Colony size: 50 to one million birds
Location: Tropical climates in Africa, Asia, Central and South America, and Europe

Living in groups helps flamingos find food and offers protection from predators. Flamingos also form breeding colonies. Females lay one or two eggs. They are cared for by both parents. When the chicks are born, they have white feathers. As they grow, the food they eat affects their feathers, turning them a shade of pink.

Flamingos march together as a group. They use sounds to communicate and warn each other of danger.

Flamingos make nests from large, cone-shaped mounds of mud.

Puffins

At a Glance

Number of species: Four—Atlantic, horned, tufted, and Rhinoceros Auklet
Location: The North Atlantic Ocean, Iceland, Greenland, Norway, Russia, France, Northeastern Canada and U.S.
Colony size: Thousands to hundreds of thousands

Puffins live most of their lives on the ocean. However, they spend spring and summer on land. They nest in colonies to increase their chances of survival and defend themselves against predators. Both parents protect and care for the chick. They hide the chicks in burrows or make nests in cliff openings.

puffin chick

A puffin chick is fed between four and ten times a day. A parent drops fish into its mouth or leaves fish at the opening of the burrow.

Puffin colonies nest in the same locations year after year.

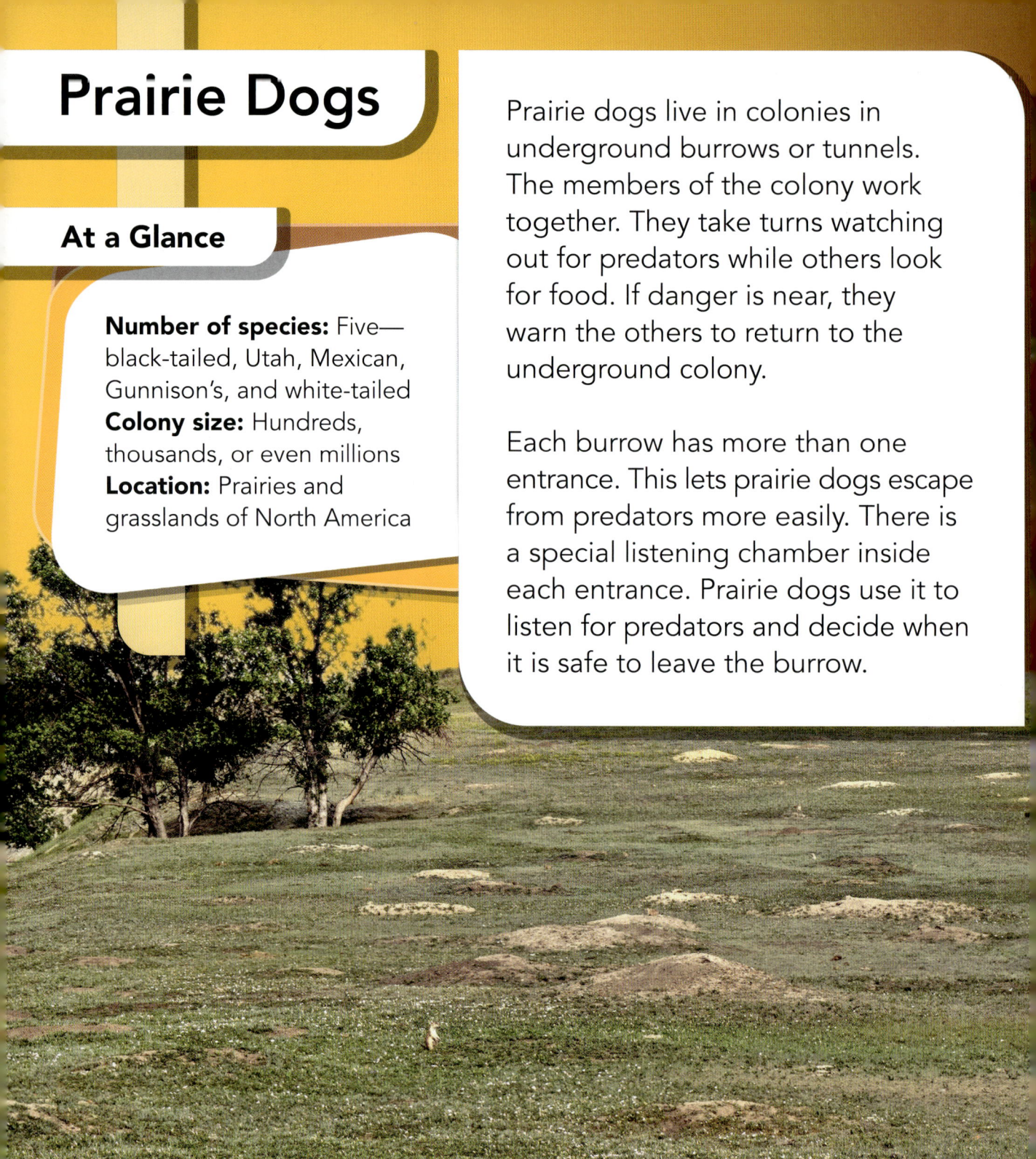

Prairie Dogs

At a Glance

Number of species: Five—black-tailed, Utah, Mexican, Gunnison's, and white-tailed
Colony size: Hundreds, thousands, or even millions
Location: Prairies and grasslands of North America

Prairie dogs live in colonies in underground burrows or tunnels. The members of the colony work together. They take turns watching out for predators while others look for food. If danger is near, they warn the others to return to the underground colony.

Each burrow has more than one entrance. This lets prairie dogs escape from predators more easily. There is a special listening chamber inside each entrance. Prairie dogs use it to listen for predators and decide when it is safe to leave the burrow.

The largest prairie dog colony ever found had more than 400 million prairie dogs!

Sticking Together

Prairie dog families include one male, several females, and their pups. They often build their burrows next to other family members, forming a group known as a coterie.

When prairie dogs greet each other, they touch their teeth together to make sure they are from the same coterie!

Prairie dog colonies are also called towns.

Town Criers

Prairie dog burrows include connected tunnels that lead to different rooms for sleeping, raising young, and even going to the bathroom!

Guards are an important part of prairie dog colonies. They help keep all of the colony safe. When a prairie dog guards its colony, it keeps watch by standing just outside its burrow. If a predator such as a coyote, hawk, or badger approaches, the prairie dog gives a warning sound. It uses a different yip, squeak, or bark for each predator.

After warning the others in its colony, the guarding prairie dog dives into the burrow to hide. The other prairie dogs hear the message and also hide.

Did You Know?

Prairie dogs got their name because of the sound they make, which is similar to a dog's bark.

This prairie dog is displaying a behavior known as a "jump-yip." It communicates to the colony that there are no predators nearby and it is safe to continue looking for food.

19

California Sea Lions

At a Glance

Number of sea lion species: Six, including California and Steller
Colony size: Hundreds to thousands
Location: The Pacific coast of North America and the Galapagos Islands

California sea lions are **social** animals. They live in large colonies, spending their time both in the ocean and on land. They form smaller groups within their colonies. These groups include one male, several females, and their pups.

California sea lions communicate with each other through loud barks, growls, and grunts—even underwater!

Females and their young make unique sounds to find each other. When a mother finds her pup, she also sniffs it to make sure the pup is hers.

California sea lions usually give birth to one pup. The mother spends just one or two days with the newborn. Then, she returns to the sea in search of food. If the pup's mother doesn't return in time to feed it, another female will feed the pup. Teamwork helps the colony survive.

California sea lions are safer from predators when they float in a group known as a raft.

Bats

At a Glance

Number of species: More than 1,400
Colony size: Dozens to millions
Location: Every continent except Antarctica

Living in a colony means that a bat can easily find a mate. Bats also care for other members of the colony through grooming, sharing food, and raising their young together. Female bats form temporary **maternity colonies** in warm, safe places away from predators.

*These bats groom each other to get rid of small **parasites** that bring disease to the colony.*

*Average maternity colonies have around 50 bats. The babies are **nurtured** in the colony for weeks before going out to find food on their own.*

Did You Know?

Bats help spread seeds that grow into more than 300 types of fruits. They also help control pests by eating a variety of insects, including mosquitoes.

Naked Mole Rats

At a Glance

Colony size: Less than 100 to almost 300
Location: Grasslands and deserts of eastern Africa

Naked mole rat colonies live underground in connected tunnels. Only the queen gives birth. Workers are responsible for raising the young, burrowing, sweeping, and cleaning. They have very poor eyesight, so tiny hairs on their skin act like feelers for traveling through the tunnels and finding food.

Most naked mole rats spend their entire lives working in a colony.

Since they don't have fur, naked mole rats huddle together to stay warm.

Honeybees

Honeybee colonies have a **caste** system. The colony is made up of one queen, many female workers, and many male bees, called drones. Each honeybee helps take care of the colony. The queen lays eggs. Worker bees look after the young, find **nectar**, and make honey. Drones mate with the queen.

At a Glance

Number of subspecies: More than 20
Colony size: 25,000 to 100,000
Location: Every continent except Antarctica

Sticking Together

In winter, worker bees huddle around the queen. They shake and wiggle to keep the queen warm. They rotate constantly, feeding and wiggling until warmer temperatures return.

Honeybees communicate with each other about the best places to find nectar. When a worker bee returns to the hive, it performs a dance with a series of turns and straight lines. The dance tells other workers where to find the nectar.

Some honeybee colonies live in nests in tree trunks and stumps. Other colonies live in hives. Hives are human-made structures that let humans collect honey to eat.

Ants

At a Glance

Number of species: More than 12,000
Colony size: A few dozen to thousands or millions
Location: All over the world, except for Greenland, Iceland, and Antarctica

Ant colonies include one queen and many workers. Each ant plays an important role to help the colony survive. The queen's only job is to lay eggs. Worker ants take care of the queen, look after the young, defend the nest, find food, and even take out the trash! Some ant species have even formed supercolonies of many nests and multiple queens. Supercolonies can have millions or billions of ants!

Worker ants remove dead leaves, ants, and other waste from the nest.

Ants communicate messages about predators or where to find food. They touch each other with their antennas or move their bodies in certain ways.

Sticking Together

Most ants make nests in one location. However, army ant colonies travel from place to place in search of food. When the queen is ready to lay eggs, worker ants surround her, protecting her until the eggs hatch.

These army ants have formed a protective colony around the queen and her eggs.

Green Sea Turtles

At a Glance

Number of sea turtle species: Seven, including leatherback and green
Colony size: Hundreds to thousands
Location: Mostly warm ocean waters worldwide

Female green sea turtles nest in colonies on the same beaches where they hatched as babies. Each sea turtle can lay more than 100 eggs! With so many eggs in the nesting colony, it is difficult for predators to find all of them. This means more of the baby turtles can make it to the ocean and survive.

Sticking Together

When baby sea turtles hatch, they leave the nest together, usually at night. They push across the sand as fast as they can to reach the water before predators such as crabs, seagulls, and raccoons attack them.

baby green sea turtle

green sea turtle eggs

Portuguese Man O'Wars

At a Glance

Location:
Tropical oceans
Colony size:
Four to hundreds or thousands

A Portuguese man o'war looks like a single creature. However, it is actually a colony of individuals called **clones**. Each part of the man o'war has a job to do, such as hunting and feeding.

Portuguese man o'war colonies don't actually swim. Instead, they float on the ocean **currents** or get pushed along by the wind. Sometimes there can be thousands floating together at a time.

*The Portuguese man o'war colony protects itself by stinging predators with **venom** from its tentacles.*

Glossary

breed To have babies

brood pouches Flaps of skin on certain penguins for keeping eggs or chicks warm as they develop

burrows Holes or tunnels made by animals for shelter

caste A group of insects in a colony that look different and have different roles

clones Individuals in a species that share identical genetic characteristics

currents Steady movements of water in a certain direction

mammals Warm-blooded animals with backbones. Mammal babies drink milk from their mothers' bodies.

mate One of a pair that makes babies

maternity colonies Groups of female bats that gather in dark, quiet places to give birth and feed their young

migrate Of animals, move from one place to another, according to seasons

nectar A sweet liquid found in many flowers that bees use to make honey

nest To build or use a nest

nurtured Fed, protected, and taken care of

parasites Living things that live and feed on other living things

predators Animals that hunt other animals for food

social Describes animals that gather, communicate, and work together

species A group of living things with similar characteristics, in which two animals can have babies

subspecies Two or more distinct categories of one species

temporary For a short period of time

venom A poisonous substance produced by some animals

Learning More

On the Web

www.natgeokids.com/uk/discover/animals/insects/ant-facts
Find out mind-blowing facts about ants, including interesting species such as fire ants and bullet ants.

www.pbs.org/wnet/nature/blog/penguin-fact-sheet
Want to learn about penguin behavior, habitats, and more? Check out this site for facts about these flightless bird colonies.

https://nationalzoo.si.edu/animals/naked-mole-rat
Go to the Smithsonian National Zoo's website to see videos and find fun facts about the world's longest-living rodent: the naked mole rat.

Books

Duling, Kaitlyn. *Animals of the Grasslands: Prairie Dogs*. Bellwether Media, Inc., 2020.

Daly, Ruth. *Bringing Back the Lesser Long-Nosed Bat*. Crabtree Publishing, 2020.

Winters, Kari-Lynn. *Buzz About Bees*. Fitzhenry and Whiteside, 2013.

Index

Africa 8, 14, 23

Antarctica 5, 8, 11, 22, 24, 26

Asia 14

Australia 8

breeding 4, 5, 6, 8, 14

burrows 9, 15, 16, 17, 18, 23

Central America 14

clones 29

communication 14, 19, 20, 25, 26

Europe 14

maternity colonies 22

migration 7, 13

nesting 5, 6, 7, 8, 9, 12, 14, 15, 25, 26, 27, 28

New Zealand 8

North America 15, 16, 20

predators 4, 7, 10, 12–16, 18, 19, 21, 22, 26, 28, 29

South America 8, 14, 20

About the Author

Kylie Burns is an author and teacher who has always been fascinated by nature. She has written more than two dozen books for children on a variety of topics including animal life cycles, sports, famous celebrities, STEM science, and more. Her own little colony includes her husband, three children, and one very bossy dog!